ISKANDER

Mobile Tactical Aero-Ballistic/Cruise Missile Complex

HUGH HARKINS

Copyright © 2017 Hugh Harkins

All rights reserved.

ISBN: 1545053448
ISBN-13: 978-1545053447

Iskander

Mobile Tactical Aero-Ballistic/Cruise Missile Complex

© Hugh Harkins 2017

Createspace Independent Publishing Platform
United States

ISBN: 1545053448
ISBN 13: 979-1545053447

This volume first published by Createspace Independent Publishing Platform for Centurion Publishing in 2017

The Author is identified as the copyright holder of this work under sections 77 and 78 of the Copyright Designs and Patents Act 1988

Cover design © Createspace Independent Publishing Platform & Centurion Publishing

Page layout, concept and design © Createspace Independent Publishing Platform & Centurion Publishing

All rights reserved. No part of this publication may be reproduced, stored in a retrieval system, transmitted in any form, or by any means, electronic, mechanical or photocopied, recorded or otherwise, without the written permission of the publishers

The publishers and author would like to thank all organisations and services for their assistance and contributions in the preparation of this volume: JSC Research and Production Corporation Konstruktorskoye Byuro Mashynostroyeniya (KBM); Volat; Rostec Corporation; Kamaz; S.P. Korolev Rocket and Space Corporation Energia; JSC MIC Mashinostroyenia (Joint Stock Company Military Industrial Corporation Scientific and Production Machine Building Association); JSC NPO Energomash, the Ministry of Defense of the Russian Federation, Lockheed Martin and the United States Department of Defence

CONTENTS

	INTRODUCTION	i
1	TOCHKA-U AND THE ROAD TO ISKANDER	3
2	ISKANDER-M/E COMPLEX	15
3	SERVICE INTRODUCTION AND DEPLOYMENT	45
4	GLOSSARY	59

INTRODUCTION

The Iskander (Искандер) mobile tactical aero-ballistic/cruise missile complex has been developed for domestic Russian service, Iskander-M, and the export market, Iskander-E, as a survivable 21st century replacement for older generation tactical ballistic missile complex's dating back to the Soviet era.

This volume sets out to detail the Iskander-M/E missile complex in service with the Russian Federation Land Force and export customer(s). All technical data relating to the respective weapon systems and their components have been provided by the respective design bureau/offices, as has much of the imagery and graphics with additional impute from the Ministry of Defense of the Russian Federation, the United States defense industry and Department of Defense.

1

TOCHKA-U AND THE ROAD TO ISKANDER

The Iskander (Искандер) high-precision tactical aero-ballistic/cruise missile, often referred to as TBM (Tactical Ballistic Missile) complex has been developed as a modern survivable TBM system and fielded by the Russian Federation as a partial counter to what that nation perceives as a growing threat from the NATO (North Atlantic Treaty Organization) alliance that, since the dissolution of the Soviet Union in December 1991, has continually advanced ever eastward against Russia's western borders and those of her allies.

While it is clear that the Iskander-M/E (apparent NATO reporting designation/name SS-26 'Stone') is a completely new design, it builds on the design legacy of previous generation TBM's such as the TOCHKA SRTBM (Short-Range Tactical (Theatre) Ballistic Missile) complex (NATO reporting designation SS-21) and the OKA (NATO reporting designation SS-23) Theatre Ballistic Missile System. TOCHKA and OKA had themselves been developed as replacements for such first generation missile systems as the FROG-7 (Free Rocket over Ground 7) and R-11A (NATO reporting name 'Scud') respectively. The Divisional level TOCHKA had a maximum firing range of 120 km, some 50 km greater than that of the FROG-7, whilst the OKA had a firing range in the region of 500 km, some 300 km greater than the 300 km firing range of the R-11A, this latter missile being deployed in missile brigades at Army level.

When introduced to service with Soviet ground forces in the 1970's (initially the TOCHKA 'Point', followed by the TOCHKA-U), the TOCHKA missile complex was among, if not the most, advanced SRTBM system in existence. TOCHKA featured high-accuracy and was highly mobile, reducing its vulnerability to counter-action. The OKA LRTBM (Long-Range Tactical (Theatre) Ballistic Missile) was introduced to service with Soviet ground forces in 1980. This advanced, for the time, highly accurate nuclear armed missile system was sacrificed as part of the Soviet-US (United States) INF (Intermediate Nuclear Forces) treaty, signed by Soviet Premier Mikhail Gorbachev and US President Ronald Regan in 1987. The INF treaty called on both parties to eliminate all intermediate and some shorter-range ballistic missiles, the Soviet Union having by far the largest number of such systems in service. In compliance of the treaty, the Soviet Union began removing the OKA system from operational service in 1987, this being completed in 1989, although it appears that a small number of systems were retained for a time in a non-nuclear role. There were of course other types of theatre missiles such as the SS-12 'Scaleboard', which had a much greater range than TOCHKA and OKA, this being in the region of 900 km.

The Iskander-M TBM forms the cutting edge of the Russian Federation modernization of its tactical surface to surface missile forces. Rostec Corporation

In regards to the aforementioned NATO expansion, this, it is clear to the Russian civilian and military leadership, is still ongoing (in 2017) as NATO looks ever further eastward to Ukraine in a drive that Russia views as a clear threat to its security. Such alarm is understandable when rhetoric language coming from some of NATO's newest member states, in particular the former Soviet Republic of Estonia, is considered. This clearly threatening language rings to the chime that Russia's very existence as a state is considered a threat and nothing short of the removal of her existence as a state would remove that threat. There are even falsifications from the Estonian Foreign Minister "they were Russian troops and they were invading us" he states referring to a Russian invasion of Estonia in 1991 that did not actually happen, this being Soviet forces moving on Soviet territory. The Estonian anti-Russian rhetoric is being ramped up in the second decade of the 21st century, with pamphlets being circulated amongst the population giving advice on what to do in the event of a *RUSSIAN* invasion despite no threat being leveled. On the contrary, the political/military actions of the Baltic States are clearly pushing a military threat against Russia's borders, increasing the very small possibility of direct confrontation.

It is such clearly hostile language from a neighbor state, which has not been threatened and is a member of the largest peacetime military alliance in human history, that has been a driving force behind Russia's modernization and rearmament of her armed forces in a push to field modern weapon system to which she could counter the perceived NATO threat on several fronts. Not least of these perceived threats has been the build-up of NATO multinational forces in Poland and the Baltic states in an effort to isolate the Russian Federations western bastion of Kaliningrad on the Baltic coast. Indeed, Kaliningrad has been the stage for the theatrics that have accompanied increasingly hostile western press reporting about Russian defensive activities on her own territory. An example of this would be the Russian deployment of advanced surface to air missiles to counter the threat from NATO fully armed combat aircraft patrolling near Russia's borders. In considering Russia's actions, one should ask themselves their own reaction or feelings in regard to the situation of armed aircraft taking off from a clearly hostile neighbor state not great distances from your own national borders. It would seem prudent that any responsible government would, if able, take defensive measures against such clear threats.

What is clear is that the NATO program of shadowing Russia's air and surface vessel movements in the Baltic region have now come at a price. NATO has to conduct such operations against a state clearly willing and, for the first time in the post-Cold War period, equally clearly capable of protecting her territorial integrity. In regards to Iskander-M, Russia has deployed such missiles to Kaliningrad, along with the older TOCHKA-U, during military exercises as a demonstration of the systems mobility/transportability and, if required, to counter further NATO military such as the much vaunted US missile defence systems, near Russia's borders.

ОТР-23. Транспортное средство для ракеты

Previous page: The Iskander missile complex is the latest in a long line of Soviet/Russian tactical ballistic missile system that started with the OKB-1 (now S.P. Korolev Rocket and Space Corporation Energia) R-11A in the mid-1950's. *Energia*
This page: The Iskander-M/E complex is a direct descendant of more recent missile complex's such as the OKA. *KBM*

The TOCHKA-U was derived from the TOCHKA 'Point' that entered service with the ground forces of the Soviet Union in the 1970's.

Although deployment of the OKA missile complex was more recent in chronological terms, this system was considered an intermediate range nuclear missile complex. In regards to operational design and actual deployment role Iskander-M's most immediate predecessor should be considered the TOCHKA-U, which remained in service with Russian Federation Land Force and Naval shore units in late 2106.

The TOCHKA-U system, which, as was the case with the TOCHKA 'point', was developed under the supervision of Sergei Nepobedimy, consisted of a number of main and secondary components: the TEL (Transporter Erector Launcher) complete with one ballistic missile; TLV (Transport Loader Vehicle) with re-fire missile; TV (Transportation Vehicle); ATV (Automated Test Vehicle); MV (Maintenance Vehicle); DES (Depot Equipment Set) and training aids. The missile, which could be guided through the entire flight, was a non-separating warhead single stage vehicle powered by a solid-propellant rocket motor. The TEL launcher would provide the missile with the relevant co-ordinates of the launch position, target co-ordinates and other relevant targeting information, which would be utilized in the calculation of a flight route to target.

Russian Federation TOCHKA-U's in the launch ready position. MODRF

Loading a missile onto a TOCHKA-U TEL. MODRF

TOCHKA-U High-Precision Tactical Ballistic Missile Complex – data furnished by JSC Research and Production Corporation, KBM

TEL/TLV chassis: three-axle, cross country, amphibious
TEL weight: 18000 kg
TEL/TLV maximum speed: 60 km/h on road and 8 km/h on water
TEL/TLV fully fueled range: 650 km
Operating temperature range: from -40° to +50° C
Crew: 4
Maximum firing range: 120 km
Minimum firing range: 15 km
Missile weight: 2010 kg
Warhead weight: 480 kg
Launch preparation time from Readiness No.1: 2 minutes
Launch preparation time form travelling positon: 16 minutes

As noted above, in late 2016, TOCHKA-U remained in operational service with the Russian Federation Land Force and Naval shore forces, operating alongside Iskander-M, although it was increasingly being superseded by its more modern brethren. Training for operation of the TOCHKA-U complex was still on-going with new personnel of the Mikhailovskaya Artillery Military Academy at the Luzhsky range in the Western Military District in 2015-2016, suggesting no imminent retirement of the system. In 2015, the TOCHKA-U still formed a part of the armory of Soviet Naval shore unit elements of the Russian Baltic Fleet being deployed to the Kaliningrad enclave.

Although officially termed a TBM complex, Iskander would be more accurately termed a tactical aero-ballistic missile complex in that its warhead vehicle (the single-stage missile) is truly aero-ballistic. That said, in Iskander-M there are two types of warhead vehicle, the aero-ballistic single-stage missile and what is termed a cruise missile of which very little is known other than that payload mass is ~480 kg and that it is very maneuverable. It is unclear (in early 2017) if the cruise missile warhead would be delivered at hypersonic, supersonic, or high subsonic speeds. The aero-ballistic missile is capable of performing high agility maneuvers and deploying countermeasures throughout its flight to defeat missile defence systems of the HTK (Hit to Kill) type such as the United States Lockheed Martin PAC-3 (Patriot Advanced Capability-3) or THAAD (Theatre High-Altitude Area Defense). Such systems, the Lockheed martin description of which likens to "stopping a bullet with a bullet", are effective against non-maneuvering ballistic trajectory warheads.

Although not the main driving force behind its development, Iskander-M, even prior to its deployment with Russian Federation Land Force Missile Brigades, has been a major factor in the Russian counter to the United States tactical/theatre ballistic missile defense element of the wider National Missile Defense programs employing the aforementioned HTK technique to intercept incoming ballistic missiles. While Russia views missile defence systems deployed in Eastern Europe as a clear threat to her own security, the United States position has always been that missile defence was not aimed at countering Russian ballistic missiles, although this argument immediately lost much of its credibility when the intended missile basing, Eastern Europe, was taken into consideration. While such missile defence systems would be an ineffective counter to a full-scale strike involving large numbers of missiles, the Russian view is that the threat to any element of her nuclear deterrent capability is untenable, resulting in the decision to neutralize the perceived threat by the introduction a number of countermeasures to the American system. Foremost among these counters are EW (Electronic Warfare), air and surface launched long-range cruise missiles and, of course, Iskander-M with the high survivability inherent in its countermeasures equipped maneuverable warhead delivery vehicles.

In the anti-BMD (Ballistic Missile Defense) role the highly maneuverable warhead vehicles are designed to hit targets actively protected by long-range and point-defence surface to air missile systems. Survivability is enhanced by the maneuver profile of the warheads throughout their flight right time down to the terminal/impact phase. Although the maneuvering warhead would, of course, lose valuable kinetic energy, even in the terminal phase of the flight of the aero-ballistic vehicle, speeds of several thousand meters per second are expected to be achieved. Such high speeds, even in the terminal/impact phase, bring the system well within the accepted parameters of hypersonic flight that is generally defined as having a lower limit of ~Mach 5 (6174 km/h), this varying slightly with flight altitude.

Previous page: TOCHKA-U missile launches. This page: TOCHKA-U from the Russian Federation Baltic Fleet deployed to Kaliningrad in 2015. MODRF

Whilst the Russian view was that the deployment of missile defence elements in Eastern Europe in 2016 was an attempt to push the strategic balance in favor of NATO, deployment of systems such as Iskander-M and accelerated development of maneuverable warheads for ICBM's are designed to address, indeed redress any small-scale short term strategic advantage that missile defence may have provided.

In regards to the above mentioned development maneuverable warheads for ICBM's, this aggressive development effort on the part of Russia has been spurred by the US advances in missile defense technology. Russian developments are not surprising since it is inevitable that the introduction of any new military capability will inevitably lead to the introduction of counters to that capability by interested parties. Such warhead developments most probably build on research and development that commenced in the 1960's. Two separate orbital maneuvering warheads were developed for the UR-200/UR-500 orbital weapon programs, the AB-200 for use on the IBM UR-200 and the AB-500, which would arm the larger IBM UR-500 missile complex. Such warheads were designed in such a manner as to be capable of maneuvering in Earth's atmosphere after de-orbit to increase accuracy. To this end, OKB-52 (JSC MIC Mashinostroyenia) had developed the MP-1 vehicle which was to be capable of using aerodynamic control to maneuver in Earth atmosphere at high hypersonic speeds, this having been accomplished in what was described as a successful test in 1961.

Graphic showing engagement sequence for THAAD. While effective against ballistic profile missiles like the R-11 and their modernized derivatives, systems such as THAAD, in their current incarnation, are not designed to be effective against aero-ballistic missile systems in the class of the Iskander-M complex. Lockheed Martin

Modern systems, often termed hypersonic glide vehicles (this being one potential development course), are in the development phase. Deployment of such systems would provide an effective counter to any future wide-ranging US missile defense deployments employing current HTK technology.

Although INF had killed the OKA, LRTBD development continued in the Soviet Union and, following the breakup of the union in December 1991, the independent Russian Federation. Development was pushing toward a system with longer range than the TOCHKA, but, certainly in regards to export derivatives, shorter range than the OKA. The hard economic situation in post-Soviet Russia during the 1990's slowed development of the new system that would be referred to as Iskander. Into the 21st century, the pace of development increased, leading to orders for development/pre-production systems from the MODRF (Ministry of Defence of the Russian Federation) that allowed operational testing/development of the weapon system in the second half of the first decade of the 21st century. The first Iskander missile launch was conducted in 2007 and the missile complex achieved a limited operational capability with a few test missile complexes at Kapustin Yar in 2009. Fully operational Iskander-M Missile Brigades began to receive their equipment in 2010. Unsubstantiated reports that a number of Iskander-M complex missiles were launched during the 2008 Russo-Georgian war are here dismissed pending credible confirmation, perhaps at a later date, if indeed such operational launches took place.

Per firing missile the Iskander M/E system requires 3-5 times less personnel than the TOCHKA and 6-12 times less personnel than a 1960's vintage (including updated variants) R-11 'Scud' complex. Iskander M/E requires three time less system vehicles per launcher than the TOCHKA and five times less than the R-11.

In comparison to the 18000 kg TOCHKA three-axle, cross country, amphibious TEL, the Iskander MAZ-79306 ASTROLOG wheeled 8x8, all terrain TEL is more than twice the total operational weight. Even at the lower of both the documented maximum road speeds of 70 and 80 km/h, the Iskander TEL/TLV is faster than the 60 km/h TOCHKA TEL/TLV, allowing for quicker deployments to fire positions. The Iskander TEL/TLV were also longer-legged with a range of 1000 km compared to the 650 km of the TOCHKA TEL/TLV. Manpower requirements were also reduced to 3 crew for the Iskander TEL compared with the 4 crew for the TOCHKA TEL.

Although both missile systems have a 480 kg payload, the Iskander aero-ballistic missile, at a launch weight of 3800 kg (3.8t), is almost twice as heavy as the TOCHKA system missile, which has a launch weight of 2010 kg (2.1t). The larger Iskander complex aero-ballistic missile has a maximum/minimum firing range of 280 km for the Iskander-E (the Iskander-M aero-ballistic missile has a considerably longer firing range) which was well in excess of the 120 km range of the TOCHKA system. The TOCHKA system, however, had a considerably lower minimum firing range of 15 km compared to the 50 km of the Iskander system.

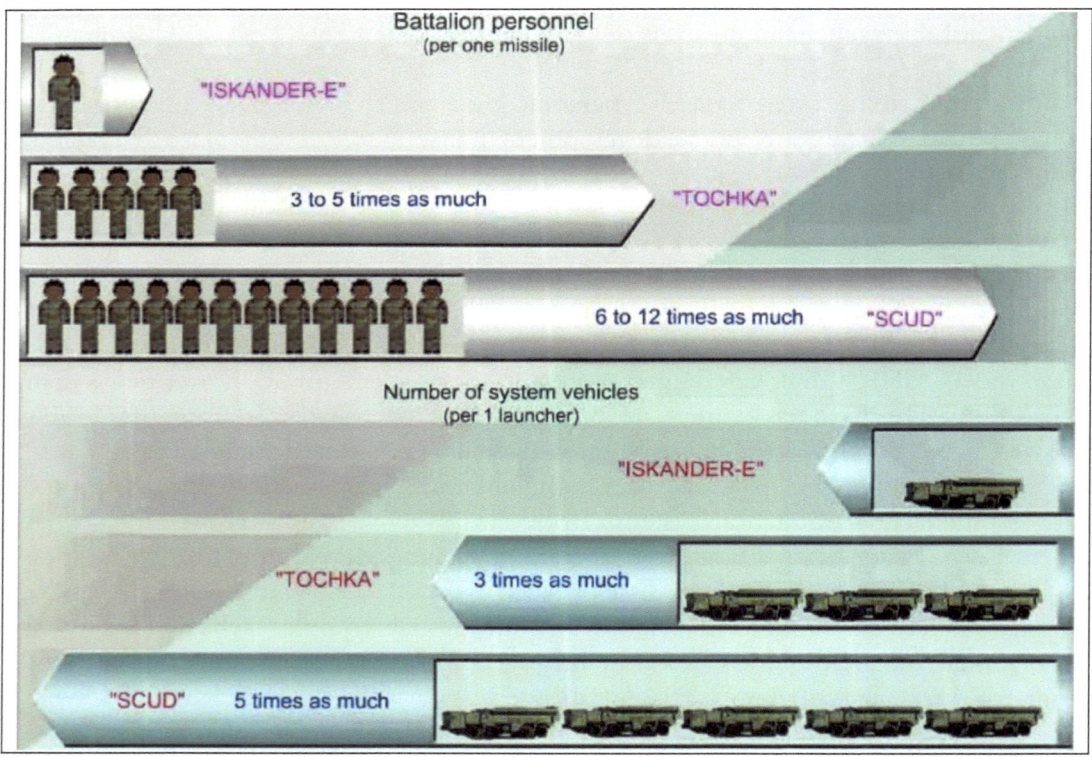

Graphic depicting the scale of reduction in terms of personnel and vehicle numbers required for the Iskander-E complex in comparison to previous generation complexes, in this case TOCHKA and 'Scud'. KBM

2

ISKANDER-M/E COMPLEX

The Iskander-M TBM (Tactical Ballistic Missile) complex has been associated with several designations, 9M723 (9K723) for the aero-ballistic missile armed complex and 9M728 (SSC-7) for the cruise missile armed complex. It has, however, proven difficult to receive qualified definitive designations from the manufacturers or customer. It is unconfirmed to what degree the aero-ballistic and cruise missile launchers are interchangeable on the TEL.

The Iskander-M missile complex comprises five major wheeled components - TEL (Transport Erector Launcher), TLV (Transport Loader Vehicle), CSV (Command and Staff Vehicle), SCMV (Scheduled Checks and Maintenance Vehicle) and the LSCV (Life Support Vehicle). There are a number of other components, not least of which are the missiles, the Depot Equipment Set and Training Aids.

The export variant of the Iskander-M, designated Iskander-E, is identical in most respects to the Russian service variant, there being few confirmed major differences other than the shorter range of the 'E' missiles. Iskander-E can be equipped with the same conventional warheads as the Iakander-M. It is speculated that the Iskander-M could be equipped with a tactical nuclear warhead, although it is stressed that there is no confirmation of this from either the primary contractor or the customer, nor is there (in early 2017) any credible evidence from third party sources to suggest the existence of such a warhead option.

Each Iskander-M Brigade set in service with the Russian Federation Land Force consists of 51 vehicles - 12 TEL, 12 TLV, 11 CSV, 14 LSCV, 1 SMCV and 1 information preparation station as well as sets of high-precision guided missiles, depot equipment and training aids. The entire system is designed to be transportable by rail, sea and air with an assigned service lifespan of 10 years.

The Volat MZKT-7930-300, selected as the chassis for the missile handing elements of the Iskander system - the TEL and TLV, was developed from the MZKT-7930. This latter chassis was used for a number of applications including the Bal-E Bereg and Bastion anti-ship cruise missile/coastal defence complexes and the antenna post for the S-400 long-range surface to air missile complex.

Iskander-M TEL with both aero-ballistic missiles in the elevated launch position. KBM

The basic operating parameters and specifications of the MZKT-7930-300 chassis are laid out below.

MZKT-7930-300 Performance – data furnished by Volat
Wheel drive: 8x8 **Max speed:** 70 km/h **Gradeability:** 55% **Angle of approach:** 25 deg. **Angle of departure:** 35 deg. **Ground clearance:** 400 mm **Climbing steps:** up to 350 mm **Side slope:** 40% **Turning radius (curb to curb):** 15 m **Fording depth with pas:** 1.4 m **Trench width:** 2 m **Cruising range:** 1000 km **Operating temperature:** -45° to +50° C **Fuel tank capacity:** 3 x 385 litre **Air transportable:** can be carried in Antonov An-22 'Cock' and An-124 'Ruslan' military transport aircraft in service with the Russian Federation Aerospace Forces

The Volat MZKT-7930-300 chassis formed the basis of the Iskander-M MAZ-79306 ASTROLOG TEL vehicles. KBM/Volat

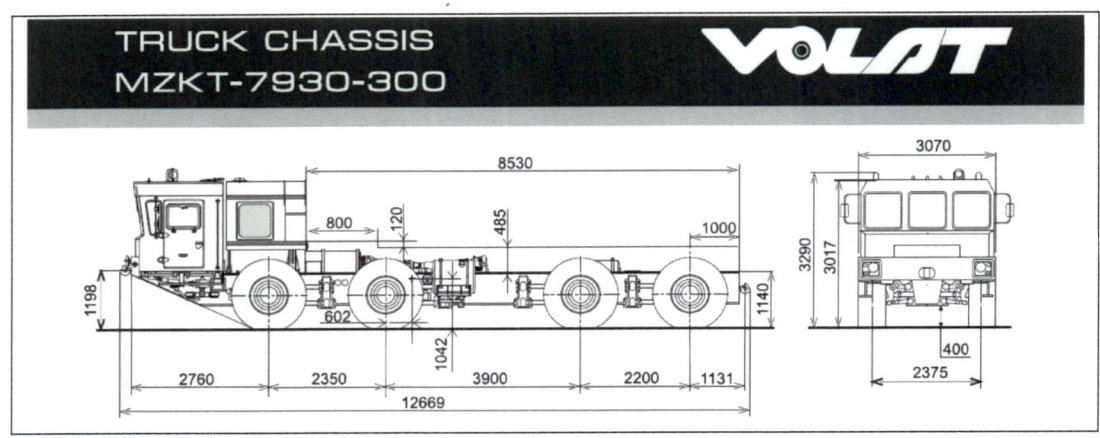

Graphic depicting the overall dimensions of the MZKT-7930-300 chassis used in the Islander TEL/TLV. Values are given in mm. Volat

MZKT-7930-300 Characteristics - data furnished by Volat

Weights
Gross vehicle weight: 45200 kg
Curb Vehicle weight: 20500 kg
Payload: 24200 kg
Permissible weight on axle
1st and 2nd axles: 2 x 11050 kg
3rd and 4th axles: 2 x 11550 kg
Engine & Transmission
Type: YAMZ-846 diesel V-shape, 8-cylinder
Power: 370 kW (503 h.p.) at 2100 min^{-1}
Maximum torque: 1960 Nm (200 kg/cm)
Transmission type: MZKT automatic
Number of gears: 6 forward and 1 reverse
Transfer case: 2-speed with lockable inter-bogie differential
Gear ratio: 1:1.00/1:1.601
Driving Axles, Suspension, Steering System, Braking system, Tyres and Frame
Driving axles: Central reduction gears and inter-axle and cross-axle differentials lock, planetary wheel hub reduction
First & second axles: Steerable, driven
Suspension: Independent torsional wheel suspension, heavy-duty shock absorbers
Steering system: Left/right hand drive, hydraulic power assisted
Brake system: Duel-circuit with pneumo-hydraulic drive
Tyres: 1500x600-635, CTIS
Frame: Low torsion ladder frame, coupled with bolted cross-members, steel bumper. Frame equipped in front with two towing forks and rear towing hook

MZKT-7930-300 Characteristics - data furnished by Volat (continued from page 18)

Cab & Electrical Systems
Cab: fibre-glass, 2 door, 3 seat, adjustable steering wheel, lighting instrument panel, interior lighter and reading lamp
Electrical Equipment
Nominal voltage: 24 v
Batteries: 4 pcs, 380 Ah
Single-wire, screened
Options
Cable winch as recovery or self-recovery
Powerful heater/ventilator
Air conditioner
Auxiliary heater
2 large central LCD [liquid crystal display]
Run flat types
Armoured cab

During trials the MZKT-7930-300 chassis of the MAZ 79306 ATROLOG Transporter Erector Launcher and Transport Loader Vehicle's cross country capability was demonstrated. The MZKT-7930-300 chassis, on occasions, had to have around one tone of mud removed during such trials. Volat

ISKANDER

Page 20: Three views of a computer generated 3-D model of an n Iskander-M TEL with aero-ballistic missiles. MODRF. Page 21: An Iskander-M TEL during the procedure of cycling the protective side/upper doors during the transition to launch mode. KBM **Above: Iskander-M TEL**. Volat

ISKANDER MAIN COMPONENTS

TRANSPORTER ERECTOR LAUNCHER – TEL - data furnished by JSC Research and Production Corporation, KBM

Chassis: wheeled 8x8, all terrain (MAZ-79306 ASTROLOG)
Full weight, including crew and load: 42300 kg
Maximum speed
On road surface: 70 km/h (conflicting KBM documentation states 80 km/h, the lower value being confirmed by Volat)
On unpaved ground: 40 km/h
Fully fuel cruise range: 1000 km
Number of missiles on launchers: 2
Time to launch first missile from highest readiness state: No more than 4 minutes
Interval between missile launches: up to 1 minute
Operational temperature range: up to ± 50° C day or night in all-weather (Volat documentation states -45° to +50° C)
Crew: 3

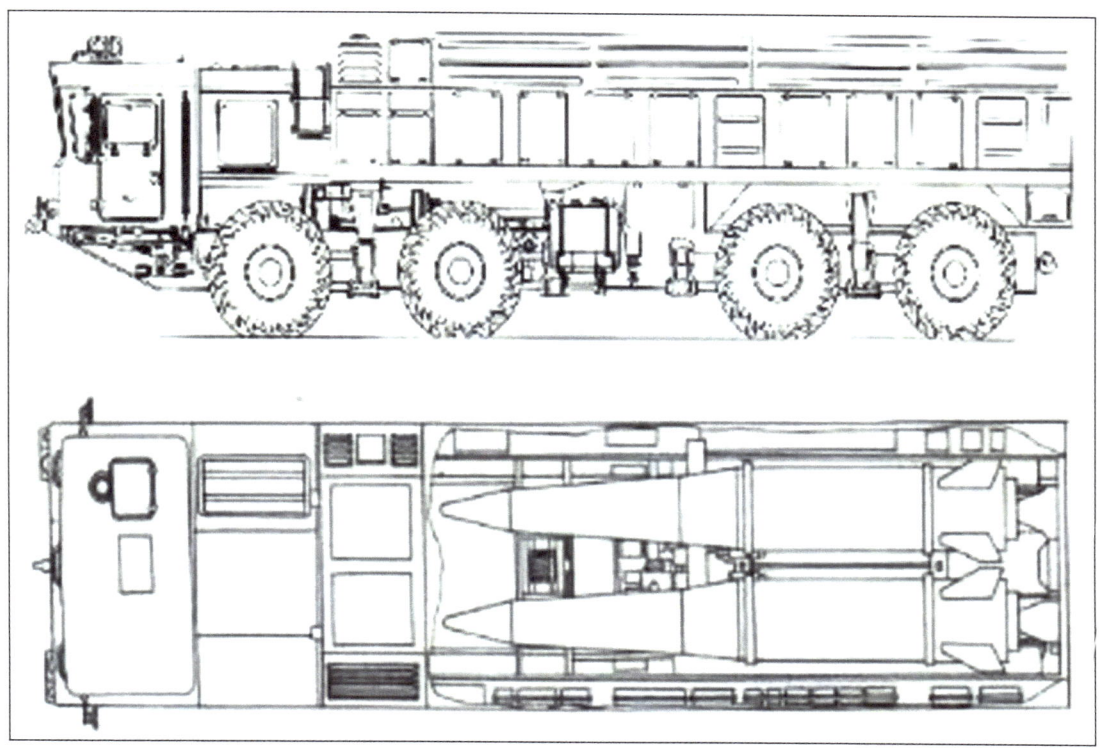

Iskander-M TEL, with aero-ballistic missiles, general arrangement (top) and general layout graphic with missiles closed off (above). KBM

Iskander-M TEL equipped with cruise missile launchers (top) and with two aero-ballistic missiles in the launch ready position (above). KBM

Top: Graphic showing the general layout of the Iskander-M TLV. KBM Above: A TLV MZKT-7930-300 chassis is put through its paces during development. Volat

Transport Loader Vehicle – TLV - data furnished by JSC Research and Production Corporation, KBM

Chassis: wheeled 8x8, all terrain (MAZ-79306 ASTROLOG)
Full weight, including crew and load: 40000 kg
Maximum speed
On road: 70 km/h
On unpaved ground: 40 km/h
Fully fueled cruising range: 1000 km
Number of transported missiles: 2
Boom crane: derrick
Crew: 2

Top: **MAZ-79306 ASTROLOG.** KBM Above: The Iskander-M complex Command and Staff Vehicle, Scheduled maintenance and checks vehicle and life support vehicles were based on **KAMAZ-43118-46/65225-43** chassis hybrids. KAMAZ

Iskander-M complex Command and Staff Vehicle. KBM

COMMAND and STAFF VEHICLE – CSV - data furnished by JSC Research and Production Corporation, KBM

Chassis type: KAMAZ-43101 (with van)
Full weight of complete filled item: 14000 kg
Number of automated workstations: 4
Communications: Includes SATNAV [Satellite Navigation]
Range coverage of radio set on the move
USW: up to 25 km
Short waves: up to 40 km
Maximum communication distance
At rest: 250 km
Mobile: 50 km
Time for solving mission calculation tasks: 10 seconds maximum
Crew: 3
Maximum time of command transfer: 15 seconds
Time of deployment/closing-in: up to 30 minutes
Operational life of system components: 10 years (includes 3 years in the field)
Operational temperature range: ±50° C

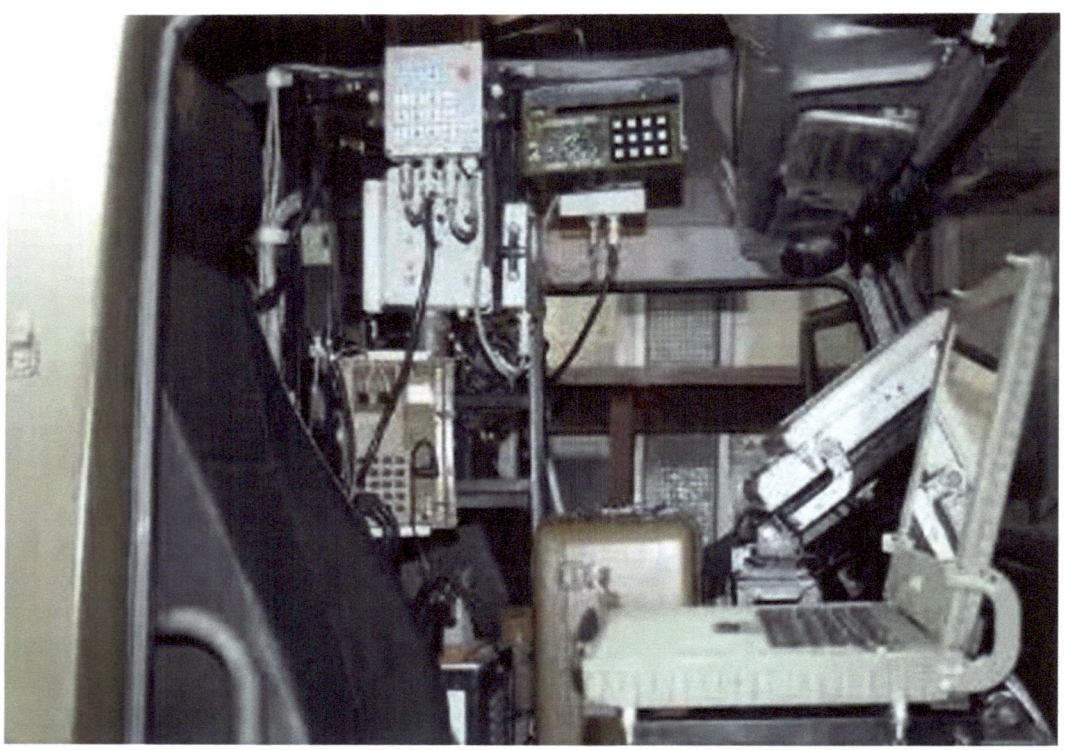

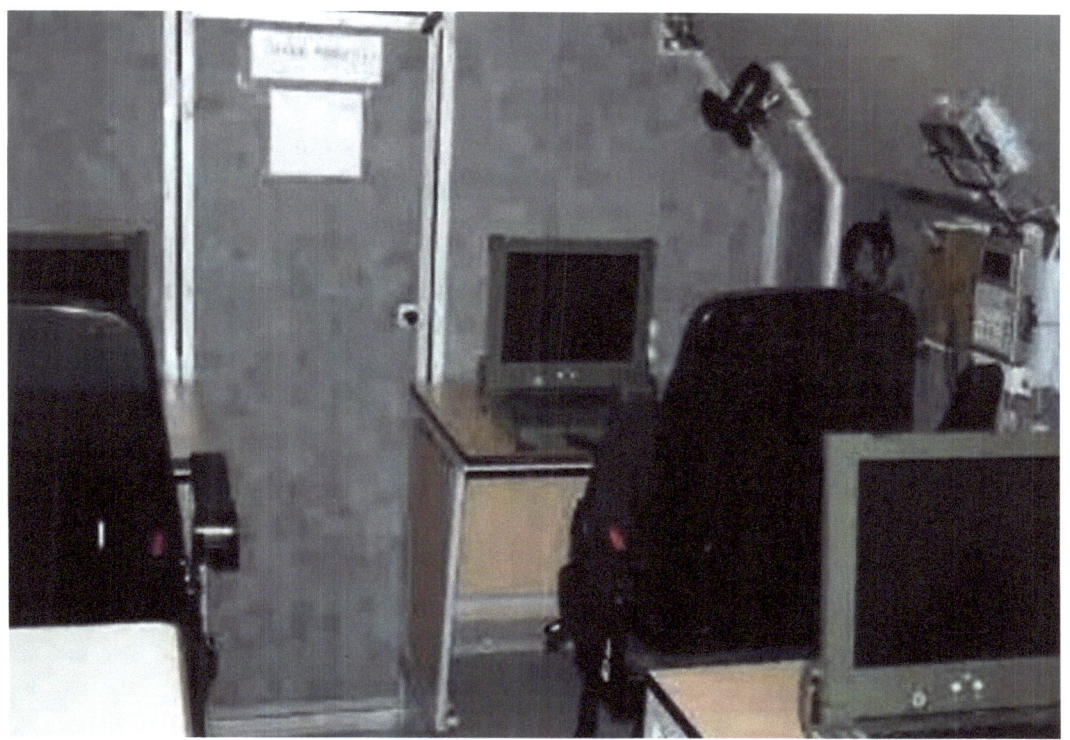

Command and Staff Vehicle interior. KBM

Scheduled Checks and Maintenance Vehicle – SCMV - data furnished by JSC Research and Production Corporation, KBM

Chassis: Kamaz type (with van)
Weight: 13500 kg
Deployment time: up to 20 minutes
Time of automated cycle of missile on-board scheduled check: 18 minutes
Crew: 3

Previous page and this page top: Iskander-M complex SCMV (Scheduled Checks and Maintenance Vehicle). Above: Graphic depicting the SCMV and Depot equipment. *KBM*

Life Support Vehicle – LSV - data furnished by JSC Research and Production Corporation, KBM

The LSV (Life Support Vehicle), using the same Kamaz type chassis (with van) as the SCMV, provides basic living facilities for the Iskander complex crew. This would include berthing and limited medical facilities.

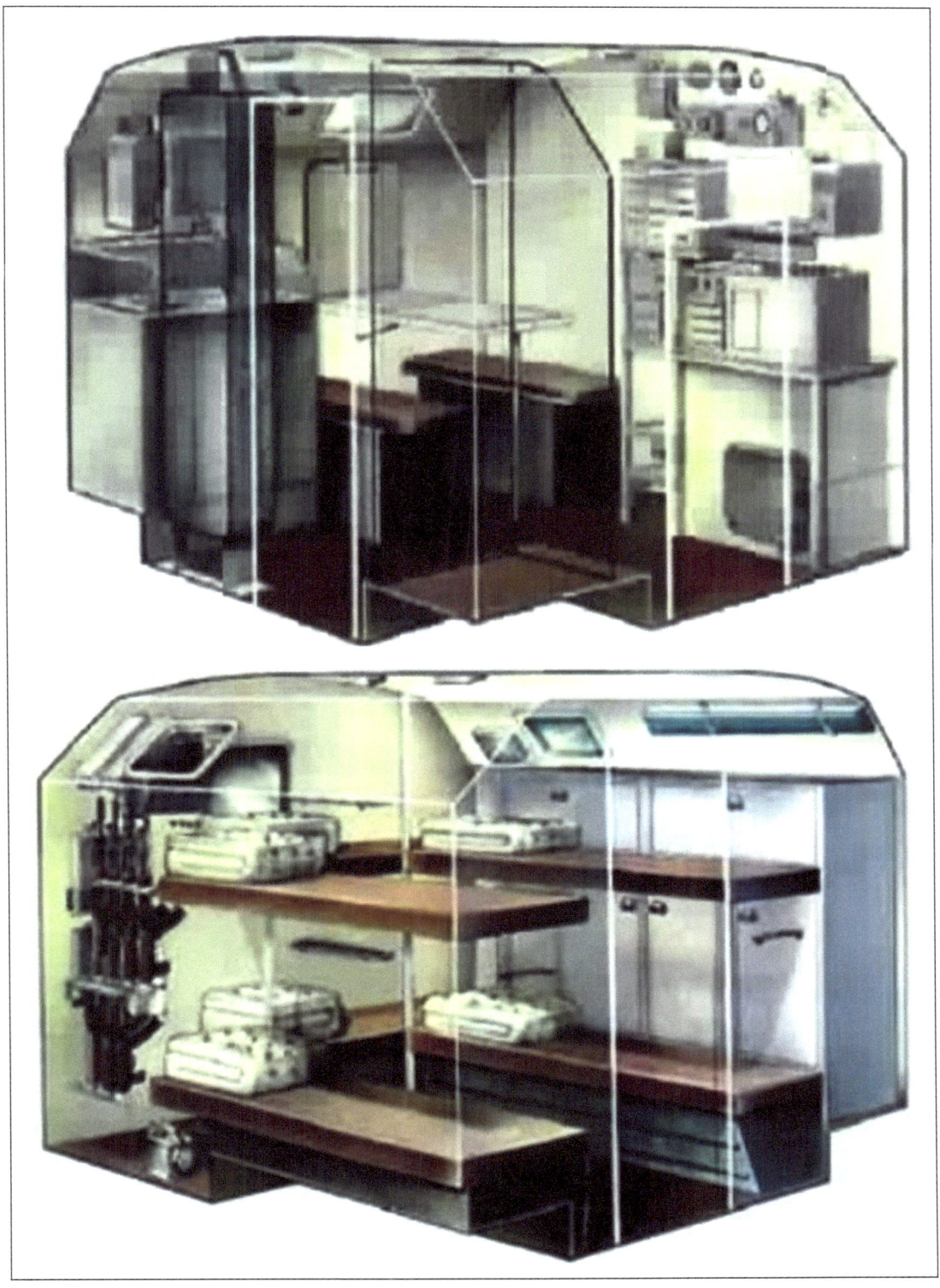

Previous page: Iskander-M complex LSV (Life Support Vehicle). This page: Interior accommodation layout of the LSV. KBM

Aero-ballistic Missile - data, which pertains to Iskander-E, furnished by JSC Research and Production Corporation, KBM

Length: 7200 mm
Maximum diameter: 950 mm
Launch weight: 3800 kg (3.8t)
Combat payload (warhead) weight: 480 kg
Launcher weight with missiles: 42300 kg
Warhead types: fragmentation, blast & fragmentation and penetrating HE
Minimum effective launch range: 50 km
Maximum effective launch range: 280 km
Missile engine: Solid propellant
Control system: strapdown inertial, integratable with optical seeker and cosmic navigational systems GLONASS and GPS

There has been much speculation as to the firing range of the Iskander-M. No official figures have been released other than confirmation of the 280 km firing range for the export Iskander-E, a number estimates putting the Iskander-M firing range at 400-500 km. However, it would appear that these estimates fall considerably short of the true value. A June 2015 Ministry of Defense of the Russian Federation release cited the commander of the Ulan-Ude based 103rd Missile Brigade, then reequipping, with Iskander-M from the TOCHKA-U complex, stating in a report to the Russian Defense Minister, General of the Army, Sergei Shoigu, that "The brigade firing capabilities will raise by two times and the firing range will be increased by five times after the rearmament". It is known that the TOCHKA-U has a maximum firing range of 120 km. The 120 km multiplied by five gives a value of 600 km, suggesting that the Iskander-M complex has a maximum firing range in that order.

Previous page: An Iskander-M complex aero-ballistic missile in the KBM stand at a trade exhibition.

A series of two stills showing an Iskander-M TEL in the launch ready configuration (top) and just after commencement of launch of an aero-ballistic missile (above). KBM

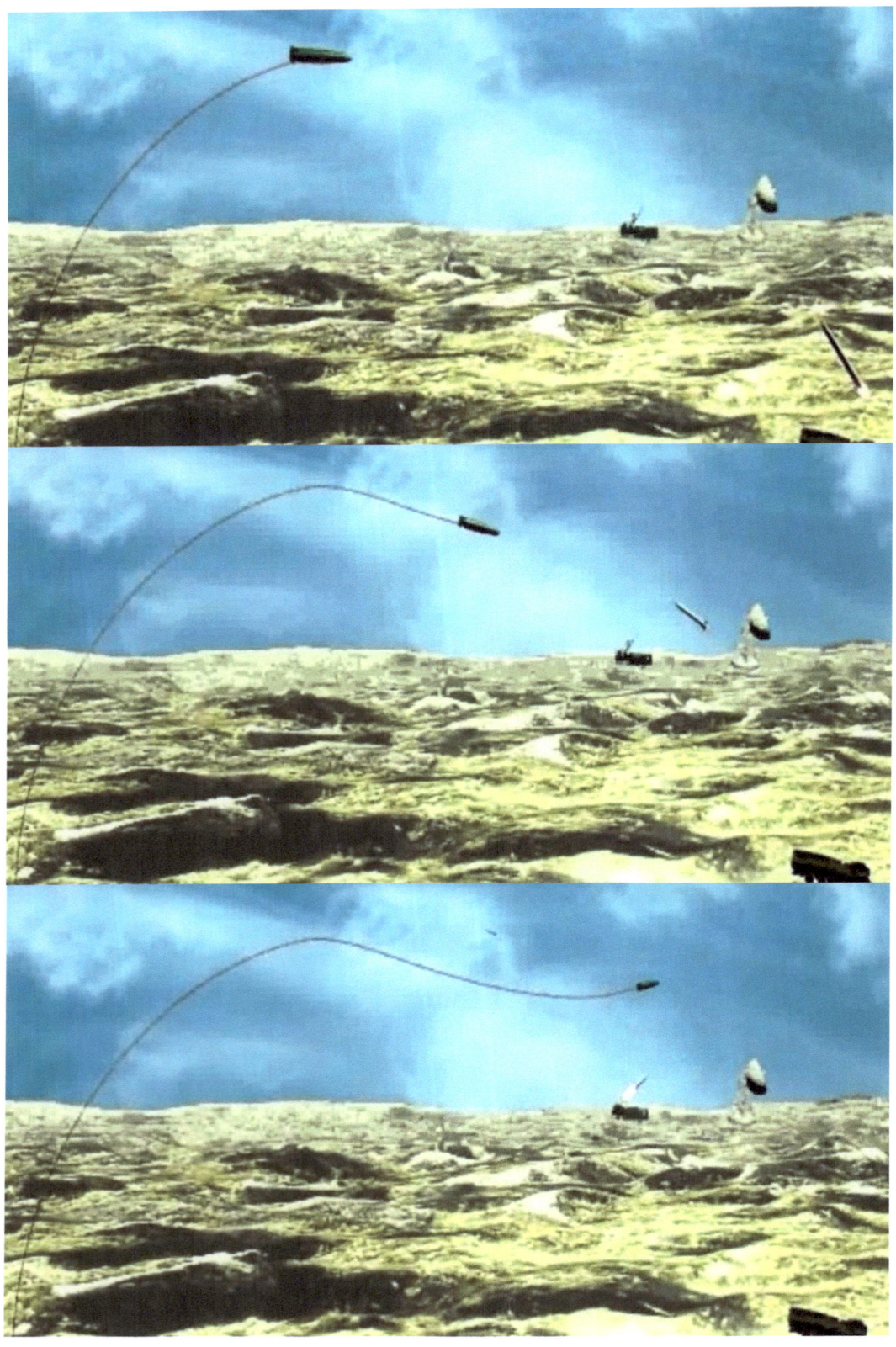

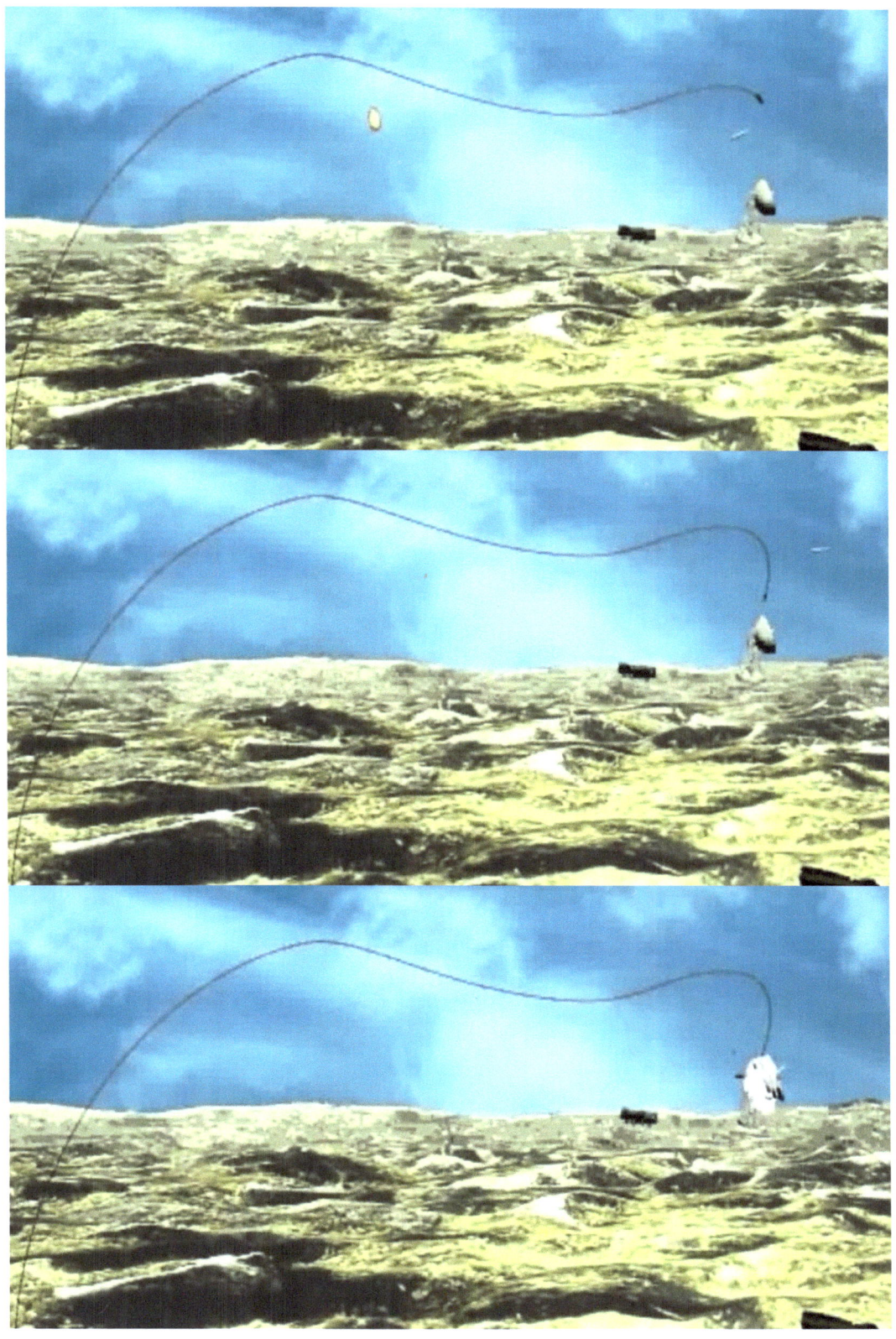

Page 36-37: A series of six graphics showing the aero-ballistic nature of the Iskander-M missile complex. The missile maneuvers to avoid a series of anti-missile defense vehicles before continuing on to strike the intended target. This page: Side-on and three-quarters front on views of Iskander-M aero-ballistic missiles. KBM

There are two known types of missile available for the Iskander-M complex, the aero-ballistic and the cruise missile; the latter spuriously referred to under the Iskander-K designation although it in fact remains part of the Iskander-M complex. Most of what is known of the complex refers to the aero-ballistic missile (previous page). MODRF No hard facts have been released about the cruise missile other than the warhead has a mass of ~480 kg. The grainy stills above serve to show the different visible characteristics of the cruise missile compared to the aero-ballistic missile. There are no released facts concerning missile specifications such as range, or indeed operational profile, the latter expected to be a low-altitude flight similar to the Kalibre and Kh-101 surface and air launched cruise missiles in service with the Russian naval and aerospace forces respectively. KBM

Iskander-M TEL's (aero-ballistic foreground and cruise missile equipped background) in various stages of cycling the launchers to the launch position. MODRF/KBM

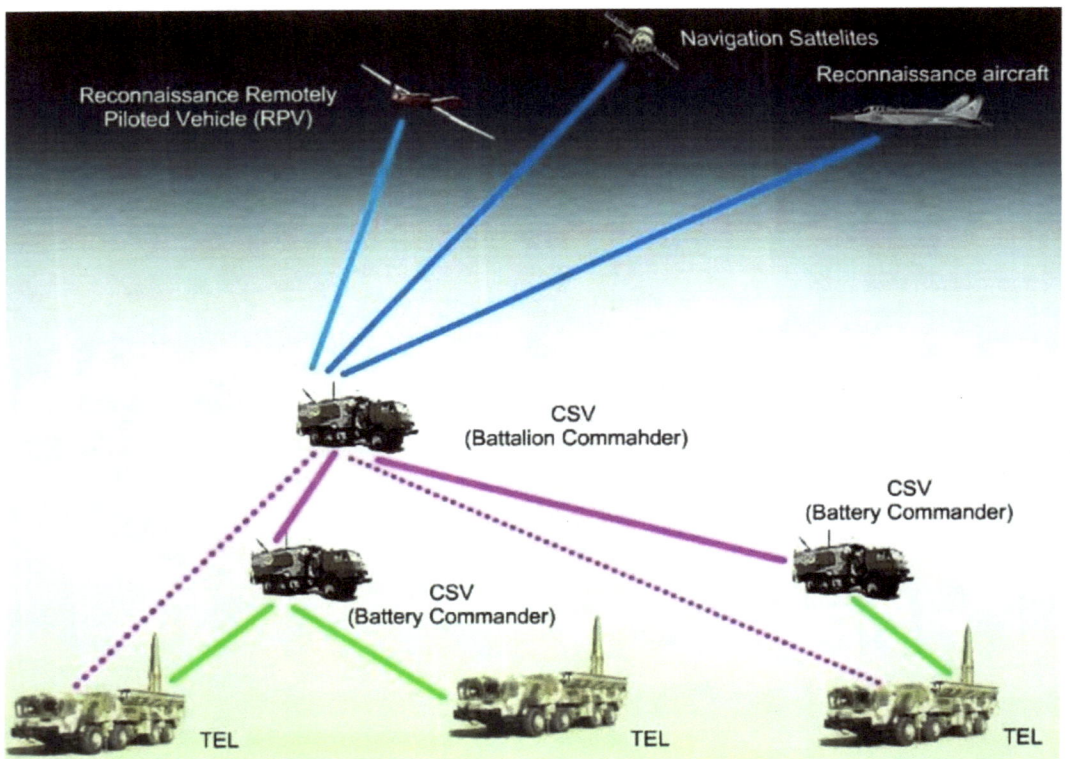

Graphic depicting the basic command and control organization for a typical Iskander-M Battalion. KBM

COMMAND & CONTROL ORGANISATION

As far as can be ascertained each Missile Brigade consists of 12 x TEL, consisting of 4 x Battalions each of 3 TEL. Targeting information from a number of sources - Navigation satellites, Reconnaissance aircraft and Reconnaissance RPV (Remotely Piloted Vehicles), also referred to as UAV (Uninhabited Air Vehicle), is downlinked to the CSV of the Battalion commander before being passed to the CSV of the Battery commanders and from here passed to the TEL, from which information goes back to the CSV Battalion commander

TARGET SETS

In theory, more or less any major target within firing range could be engaged. However, major high value target sets consist of ABM (Anti-Ballistic Missile) and air defense installation, command and control centres, MLRS (Multiple Launch Rocket Systems), Long-range artillery systems, Concentrations of personnel or equipment, Important civilian infrastructure as well, of course, as many other types of low-sized and area targets, the latter including air bases.

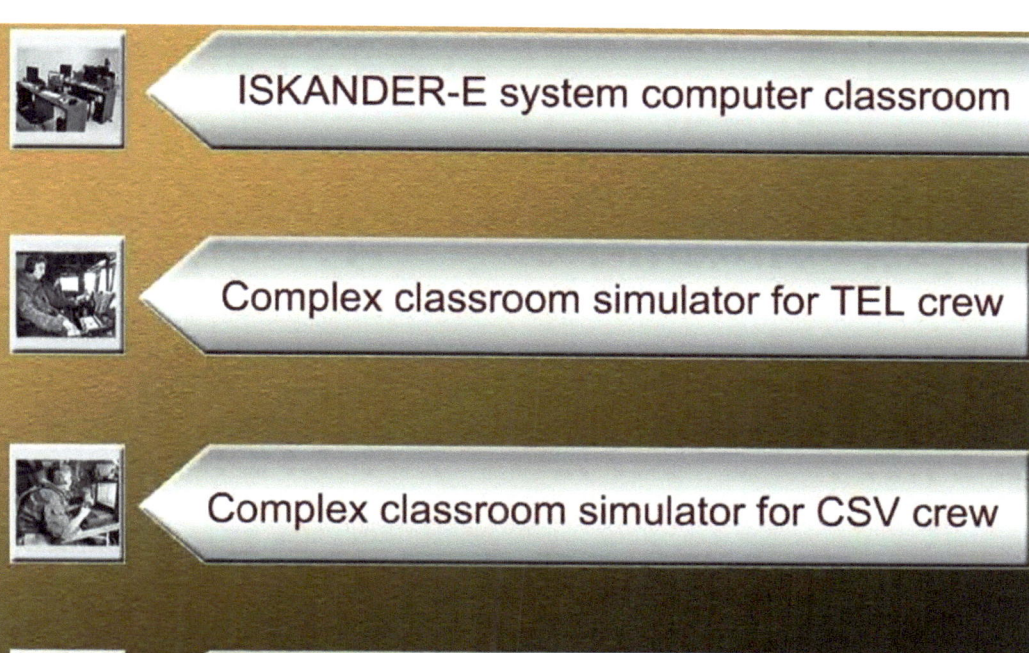

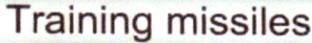

TRAINING AIDS - data furnished by JSC Research and Production Corporation, KBM

ISKANDER-M/E complex computer classroom
Complex classroom simulator for TEL crew
Complex classroom simulator for SCV crew
Training missiles

Previous page: The Iskander-M/E complex comes with an extensive non-field training element that includes various simulators and training missiles. This page: Much of the missile training is conducted in a virtual environment as shown by these rather poor quality, but relevant, graphics. KBM

3

SERVICE INTRODUCTION AND DEPLOYMENT

Iskander-M is intended to form the major element of Russian Federation Land Force short-range Missile Brigades, being considered a force multiplier in terms of operational effectiveness. As noted in chapter 1, the system is a major element of the Russian Federation counter to the United States ABM (Anti-Ballistic Missile) defense systems deployed in Eastern Europe, which, despite NATO claims that the system was not intended to counter Russian missiles, the deployment locations clearly allow them to be considered a threat to a percentage of Russian missiles in the event of a Russo/NATO conflict.

As noted in chapter 1, limited deliveries of Iskander complexes for testing and operational development commenced in 2007. However, it was not until 2010 that full-scale deliveries commenced as the Russian Federation pushed forward with a program to equip ten Missile Brigades with the Iskander-M complex. As progress was being made with the first and second Iskander-M Brigade sets previously ordered, a contract was concluded between the MODRF (Ministry of Defense of the Russian Federation) and JSC Research and Production Corporation, KBM on 4 August 2011 for delivery of a third Brigade set. This contract differed from the previous practice of contracting for individual components separately, with a new practice of contracting for an entire Brigade set of all components. This facilitated improvements in the training programs for missile brigades, which in turn increased operational availability of components of those brigades. The new practice included the mandatory training of missile brigade teams prior to the official handover of the equipment for operational use. The benefits of the new procedure for training and handover of equipment was demonstrated in 2013 when the Iskander-M set delivered that year was able to participate in a combined-arms exercise shortly after the MODRF officially accepted the handover of brigade set. All of the combat training launches conducted during this exercise were deemed successful.

On 8 May 2014, a preplanned training exercise was conducted to validate control of the Russian armed forces use of missile forces, artillery, air force and anti-aircraft defence forces in order to counter a large-scale strike of nuclear deterrence forces.

This exercise included the launch of missiles from three Iskander-M TEL (Transporter Erector Launchers). The MODRF release stated that these missiles were both aero-ballistic and cruise missile, all of which hit the designated targets. The following day, 9 May, four TEL and 4 TLV (Transport Loader Vehicles) of an Iskander-M missile brigade participated in the parade celebrating the 69th anniversary of victory in the Great Patriotic War (World War II).

Iskander-M complex TEL (top) and TLV (above) of Russian Federation newly equipped missile brigades. KBM

Iskander-M TEL's in service with the Russian Federation Land Force. MODRF/Volat

The Third Iskander-M production set was handed over to the MODRF in July 2014, equipping a missile brigade. This was followed by delivery of the fourth Iskander-M Brigade set, the second of 2014, which was delivered on 18 November that year. Handover of this latter Brigade set took place at the Combat Training Centre with the Russian army missile and artillery forces. In the month or so prior to the handover, integration and training teams worked at Kapustin Yar missile range, Astrakhan region, as the practice introduced in the 2011 predominated.

The fifth Iskander-M brigade set was delivered in July 2015, followed by the sixth in November that year. Meanwhile, training with the new missile complexes in fully combat capable missile Brigades continued apace. In the second half of June 2015, crews from an Eastern MD (Military District) Iskander-M formation completed theoretical training on the system at a facility located in the Republic of Buryatia, having previously been deployed at specialized ranges in Astrakhan region that April.

Page 48-49: Iskander-M TEL's in service with missile brigades of the Russian Federation Land Force. KBM

Elements of Iskander-M complex's in service with missile brigades of the Russian Federation Land Force. KBM/Rostec Corporation

A formation of the Southern MD conducted Iskander-M combat training launches in either late September or early October 2015 (MODRF statement was dated 1 October which is thought to be the date the exercise concluded). The missiles were launched form operational TEL's at the Kapustin Yar range. Targets sets consisted of simulated infrastructure facilities located at distances of around 300 km from the firing units. In September 2015, Eastern MD formation, based in Buryatia, conducted Iskander-M missile launches from the range in the Astrakhan region at targets some 300 km distant.

A major training exercise of 2015, Tsentre-2015 (Center-2015) included Iskander-M complexes launching cruise missiles. Another reference to the Iskander cruise missile was made in an MODRF statement issued in April 2016. This stated that an Iskander-M cruise missile variant was launched from the Kapustin Yar test range during a military exercise. The missile impacted the designated target 200 km from the launch point.

In late July 2016, the same month that the seventh Iskander-M brigade set was delivered, a Missile Brigade of the Western MD conducted exercises with Iskander-M systems at a dedicated range facility in the Leningrad region. The brigade practiced both single and multiple missile simulated launches that targeted simulated enemy command and control targets, including radar systems at simulated distances of several hundred km from the simulated launch point. This exercise also demonstrated wartime mobility of transferring from one area to another in conditions of light and radio black-out, using cross-country means to bypass public roads.

In September 2016, units of a Southern MD Missile Brigade formation, which had been activated with the Iskander-M in December 2015, conducted training with Iskander-M complexes. The formation, normally deployed in the Republic of Northern Ossetia-Alania, transferred to Kapustin Yar for the training exercises that included a number of Iskander-M electronic launches. On the 20th of the following month, Iskander-M crews from a Western MD formation conducted simulated missile launches at the Luzhsky range. The aim of this exercise was to assess crews in the preparation of missiles launches against simulated command center and radar station targets at distances up several hundred km.

The Iskander-M system is, in 2017, firmly established in Russian Federation service in both aero-ballistic and cruise missile form. In the middle of the second decade of the 21st century various contracts called for the delivery of two Iskander-M Brigade sets per year to the MODRF. This will continue until delivery of ten full Brigade sets, each with 12 TEL (a total of 24 missile launchers), is complete. There were, of course, additional launchers purchased for development/testing of the system, primarily at Kapustin Yar.

Page 53: A photo sequence of the launch and impact on target of an Iskander-M cruise missile. KBM **This page: Iskander-M TEL fire units being put through their paces during maneuvers with the Russian Federation Land Force.** MODRF/Rostec Corporation

Top: A multiple Iskander-M aero-ballistic missile launch. Above: Iskander-M had its major public debut at the 9 May 2014 parade to mark the 69th anniversary of victory in the Great Patriotic War (World War II). KBM

Page 56-57: Central Military District Iskander-M Brigade with Cruise missile TEL during winter training in early 2017. MODRF

As this is written in early 2017, there has been no credible confirmatory evidence to support assertions that Iskander-E has been purchased by and delivered to the Syrian Arab Republic. It would, however, appear that Armenia has acquired a number of Iskander-E sets, the first of which were possibly delivered in the second half of 2016, although details are sparse. Interest has also been expressed by a number of countries, notably in the Middle-East/Persian Gulf region and Russia's East European defence partner, Belarus.

Iskander-M TEL. MODRF/Rostec Corporation

GLOSSARY

AA	Anti-Aircraft
ABM	Anti-Ballistic Missile
BMD	Ballistic Missile Defence
C	Centigrade
CSV	Command and Staff Vehicle
EW	Electronic Warfare
GLONASS	Globanaya Navigoziornaya Sputnikovaya Sistema (Global Navigation Satellite System)
GPS	Global Positioning System
HE	High Explosive
HTK	Hit to Kill
ICBM	Intercontinental Ballistic Missile
INF	Intermediate Nuclear Forces (treaty signed between the Soviet Union and the United States in 1987)
JSC	Joint Stock Company
KBM	Konstruktorskoye Byuro Mashynostroyeniya;
kg	Kilogram
kg/cm	Kilogram per centimeter
km	Kilometer
km/h	Kilometer per hour
LCD	Liquid Crystal Display
LRTBM	Long-Range Tactical Ballistic Missile
LSCV	Life Support Vehicle
MD	Military District
MLRS	Multiple Launch Rocket System
MODRF	Ministry of Defense of the Russian Federation
NATO	North Atlantic Treaty Organization
PAC-3	Patriot Advanced Capability-3
SCMV	Scheduled Checks & Maintenance Vehicle
SLBM	Submarine Launched Ballistic Missile
SRTBM	Short-Range Tactical (Theatre) Ballistic Missile
TBM	Tactical Ballistic Missile
TEL	Transport Erector Launcher
THAAD	Theatre High Altitude Area Defense
TLV	Transport Loader Vehicle
US	United States
USW	Ultra-Short Wave
x	Times - multiplication
~	Approximately equal to (can also be used to mean asymptotically equal)
±	Plus or minus
°	Degree(s)

ABOUT THE AUTHOR

Hugh, a historian and author with an extensive background in astro/geophysics and studies in the wider scientific, aeronautic, astronautic and nautical technical and historical subjects, has published in excess of sixty books; non-fiction and fiction, writing under his given name as well as utilising several pseudonyms. He has also written for several international magazines, whilst his work has been used as reference for many other projects ranging from the aviation industry, international news corporations and film media to encyclopaedias, museum exhibits and the computer gaming industry. Hugh is a member of the institute of Physics, the British Geophysics Association, the European Astronomical Society and is an elected Fellow of the Royal Astronomical Society. He currently resides in his native Scotland

Other titles by the author include
Orbital/Fractional Orbit Bombardment System - The Soviet Globalnaya Raketa
Sukhoi T-50/PAK FA - Russia's 5th Generation 'Stealth' Fighter
Sukhoi Su-35S 'Flanker' E - Russia's 4++ Generation Super-Manoeuvrability Fighter
Sukhoi Su-34 'Fullback'
Sukhoi Su-30MKK/MK2/M2 - Russo Kitashiy Striker from Amur
MiG-35/D 'Fulcrum' F – Towards the Fifth Generation
Air War over Syria, Tu-160, Tu-95MS & Tu-22M3 - Cruise Missile and Bombing Strikes on Syria, November 2015-February 2016
Sukhoi Su-27SM(3)/SKM
Russian Non-Nuclear Attack Submarines – Project 877/877E/877EKM/Project 636/636.3 & Project 677/Amur 1650/950/S-1000
Russian/Soviet Aircraft Carrier & Carrier Aviation Design & Evolution Volume 1 - Seaplane Carriers, Project 71/72, Graf Zeppelin, Project 1123 ASW Cruiser & Project 1143-1143.4
Heavy Aircraft Carrying Cruiser
Light Battle Cruisers and the Second Battle of Heligoland Bight
British Battlecruisers of World War 1 - Operational Log, July 1914-June 1915
Eurofighter Typhoon - Storm over Europe
Tornado F.2/F.3 Air Defence Variant
Air to Air Missile Directory
North American F-108 Rapier - Mach 3 Interceptor
Convair YB-60 - Fort Worth Overcast
Boeing X-36 Tailless Agility Flight Research Aircraft
X-32 - The Boeing Joint Strike Fighter
X-35 - Progenitor to the F-35 Lightning II
X-45 Uninhabited Combat Air Vehicle
Into The Cauldron - The Lancaster MK.I Daylight Raid on Augsburg
Hurricane IIB Combat Log - 151 Wing RAF, North Russia 1941
RAF Meteor Jet Fighters in World War II, an Operational Log
Typhoon IA/B Combat Log - Operation Jubilee, August 1942
Defiant MK.I Combat Log - Fighter Command, May-September 1940
Blenheim MK.IF Combat Log - Fighter Command Day Fighter Sweeps/Night Interceptions, September 1939 - June 1940
Tomahawk I/II Combat Log - European Theatre, 1941-42
Fortress MK.I Combat Log - Bomber Command High Altitude Bombing Operations, July-September 1941
XF-92 - Convairs Arrow

www.ingramcontent.com/pod-product-compliance
Lightning Source LLC
Chambersburg PA
CBHW042025200526
45172CB00028B/1105